Workbook to Accompany

PLUMBING TECHNOLOGY

Design and Installation

Fourth Edition

Kevin Standiford

DELMAR
CENGAGE Learning

**Plumbing Technology: Design and
Installation Workbook, Fourth Edition**
Kevin Standiford

Vice President, Technology and Trades ABU:
David Garza

Director of Learning Solutions:
Sandy Clark

Managing Editor:
Larry Main

Senior Acquisitions Editor:
James DeVoe

Product Manager:
Mary Clyne

Marketing Director:
Deborah S. Yarnell

Marketing Manager:
Kevin Rivenberg

Marketing Specialist:
Mark Pierro

Director of Production:
Patty Stephan

Production Manager:
Stacy Masucci

Content Project Manager:
Jennifer Hanley

Art Director: B. Casey

Editorial Assistant: Tom Best

For product information and technology assistance, contact us at
Cengage Learning Customer & Sales Support, 1-800-354-9706
For permission to use material from this text or product,
submit all requests online at **www.cengage.com/permissions**
Further permissions questions can be emailed to
permissionrequest@cengage.com

ISBN-13: 978-1-4180-5092-4

ISBN-10: 1-4180-5092-X

Delmar
Executive Woods
5 Maxwell Drive
Clifton Park, NY 12065
USA

Cengage Learning is a leading provider of customized learning solutions
with office locations around the globe, including Singapore, the United
Kingdom, Australia, Mexico, Brazil, and Japan. Locate your local office at
www.cengage.com/global

Cengage Learning products are represented in Canada by
Nelson Education, Ltd.

To learn more about Delmar, visit **www.cengage.com/delmar**

Purchase any of our products at your local bookstore or at our preferred
online store **www.cengagebrain.com**

Contents

SECTION 5: Drainage, Waste, and Vent Systems

SECTION 6: System Sizing

Preface

Designed to accompany *Plumbing Technology: Design and Installation,* Fourth Edition, this Workbook provides additional review questions and exercises to challenge and reinforce the student's comprehension of the content presented in the core text.

The Workbook is organized into chapters that correspond with *Plumbing Technology: Design and Installation,* Fourth Edition. The Workbook chapters consist of:

- Objectives
- Glossary terms
- Introduction
- Review questions
- Job sheets

FEATURES OF THIS WORKBOOK

- Each chapter provides an essential study guide to the corresponding text chapter.
- The **review questions** are comprised of a variety of matching, true or false, multiple choice, short answer, and essay-style questions that are based on the materials presented in the core text and workbook.
- Each **job sheet** consists of an objective, instructions, and an activity or checklist. The job sheets range in complexity from entry level to more complex problems that require the student to perform calculations.

Together with the study tools and review questions, this Workbook's hands-on projects will help students develop practical skills and professionalism.

ACKNOWLEDGMENTS

The author and publisher wish to acknowledge the contributions of the following reviewers, who read the labs for technical accuracy:

John Breece, Red Rocks Community College, Lakewood, Colorado
Mike Roberts, St. Cloud, Minnesota

SECTION 1

Plumbing Knowledge

Plumbing Basics

OBJECTIVES

After completing this chapter, the student should be able to:

- ▪ Identify and describe the hand tools commonly used by plumbers.
- ▪ Identify and describe the power tools commonly used by plumbers.

KEY TERMS

Toolbox

Pipe wrench

Pliers

Screwdriver

No hub torque wrench

Pipe level

Hammer

Hacksaw

Chisel

INTRODUCTION

Regardless of the career path a plumber decides to take (residential, commercial or industrial), there are tools and equipment that are common to all plumbing trades. Therefore, an understanding of each tools intended use is essential to safely completing a particular task. In addition to understanding proper use, it is imperative that the plumbing student understand how to properly care for his or her tools. Maintaining plumbing tools in good operating condition will not only make completing a plumbing task easier, but it will also help prevent workplace accidents. Faulty tools are one of the leading causes of injuries in the building trades industry.

REVIEW QUESTIONS

Short Answer

1. Why is it important to use the correct tool for a specific task?

2. What are the three different career paths that a plumber can take?

3. What are the five types of drills commonly used in plumbing?

4. For what are swaging tools used?

5. For what are copper bending tools used?

True/False

6. True or False: The right tool can help a plumber complete a task more efficiently and safely.

7. True or False: The plumbing trade only uses specialty tools.

8. True or False: The tools used in the residential industry are different from the tools used for commercial plumbing and therefore if a plumber wishes to do both residential and commercial plumbing they must have two totally different sets of tools.

9. True or False: When inserting a drill bit into a drill, it is always a good practice to unplug the drill before starting.

10. True or False: Grinders are seldom used in pipefitting and are more frequently used in plumbing.

JOB SHEET 1

The Plumber's Toolbox

Name: _____ Date: _____

After completing this job sheet, the student should be able to identify the tools commonly used in all plumbing trades.

Complete the following checklist of the basic tools used by all plumbing trades.

Tool	Present	Condition
24" toolbox	☐	_____
24" pipe wrench	☐	_____
18" pipe wrench	☐	_____
10" or 12" angled jaw pliers (set of two)	☐	_____
large flat-head screwdriver	☐	_____
medium flat-head screwdriver	☐	_____
medium Phillips-head screwdriver	☐	_____
multi-head screwdriver	☐	_____
5/16" nut driver and/or no-hub torque wrench	☐	_____
24" pipe level	☐	_____
torpedo pipe level	☐	_____
claw-type hammer	☐	_____
straight cut aviation snips or a set of three	☐	_____
25'-0" retractable tape measure	☐	_____
6'-0" folding rule	☐	_____
8" or 10" adjustable wrench	☐	_____
hacksaw	☐	_____
plumb bob	☐	_____
chalk box	☐	_____
concrete chisel	☐	_____
wood chisel	☐	_____
Allen wrench set	☐	_____
copper midget tubing cutter	☐	_____
copper tubing cutter from 1/8" to 1 1/8"	☐	_____
copper tubing cutter up to 2" pipe size	☐	_____
utility knife	☐	_____
small soldering torch regulator kit	☐	_____
torch striker	☐	_____
B-tank soldering torch kit	☐	_____

Instructor's Response: Turn page

Instructor's Response

JOB SHEET 2

The Plumber's Toolbox

Name: _____   Date: _____

After completing this job sheet, the student should be able to identify the specialty tools used in all plumbing trades.

Complete the following checklist of the specialty tools used by all plumbing trades.

Tool	Present	Condition
pipe nipple extracting set	☐	_____
thread tapping tool	☐	_____
inside plastic pipe cutter	☐	_____
basin wrench	☐	_____
mini-hacksaw	☐	_____
nail puller	☐	_____
smooth-jaw pipe wrench	☐	_____
plastic pipe saw	☐	_____
needle-nose pliers	☐	_____
locking pliers	☐	_____
6" combination pliers	☐	_____
copper flaring tool	☐	_____
copper tubing bender	☐	_____
flexible tubing cutter	☐	_____
copper swaging tool	☐	_____
basket strainer tool or internal wrench	☐	_____
multi-purpose knife/pliers tool	☐	_____
carpenter square	☐	_____
metal stud punch	☐	_____
strap pipe wrench	☐	_____
chain pipe wrench	☐	_____
miniature hacksaw	☐	_____
wallboard saw	☐	_____
flexible tubing crimping tool	☐	_____
ball pein hammer	☐	_____
inside plastic pipe cutter	☐	_____
cast-iron chain cutters	☐	_____
internal cast-iron cutters	☐	_____

Instructor's Response: Turn page

Instructor's Response

Safety

OBJECTIVES

After completing this chapter, the student should be able to:

- Identify and select the proper fire extinguisher for a particular fire.
- Understand basic OSHA regulations.

KEY TERMS

OSHA

Safety

First aid

Electrical shock

Electrical injury

Electrical safety

Material handling

Affix

Casters

Conduct

Lifts

PPE

INTRODUCTION

Completing an assigned task on time and on budget is a priority of every plumbing manager. It should also be a priority of everyone associated with a project. However, the most important aspect of any project is safety. Completing a project on time is pointless if doing so produces an injury or fatality. To help ensure that safety is the priority the federal government has established the Occupational Safety and Health Administration (OSHA). The Occupational Safety and Health Administration was established within the Department of Labor and was authorized to regulate health and safety conditions for all employers with few exceptions.

REVIEW QUESTIONS

True/False

1. True or False: Good safety training can minimize injury and fatal accidents.

2. True or False: OSHA does not regulate toxic and hazardous substances.

3. True or False: OSHA mandates that companies using dangerous products have a hazard communications program.

4. True or False: Always ensure that the lanyard is routed so that it will not tangle around arms, legs, or neck if a fall occurs.

5. True or False: When working around electricity, it is OK to use a metal ladder as long as the ladder is properly grounded.

Short Answer

6. Define the following terms:

   a. GFCI

   b. Hazardous substance

   c. Affix

   d. PPE

7. How does OSHA define a working height?

8. What are the three common types of motorized manlifts used?

9. How does OSHA define a confined space?

10. How does OSHA define a trench?

JOB SHEET 1

Personal Protection Equipment

Name: _____   Date: _____

After completing this job sheet, the student should be able to identify the personal protection equipment (PPE) commonly used in the plumbing trade.

Complete the following checklist of the PPE commonly used in the plumbing trade.

Tool	Present	Condition
safety glasses/goggles	☐	_____
face shield	☐	_____
hard hat	☐	_____
boots/shoes	☐	_____
gloves	☐	_____
knee pads	☐	_____
ear plugs and muffs	☐	_____
special clothing based on threat	☐	_____
harness, lifeline, and lanyard	☐	_____
fire extinguisher	☐	_____
first aid kit	☐	_____
dust mask/respirator	☐	_____
ground fault circuit interrupter	☐	_____
specific eye and/or face shield or goggles	☐	_____

Instructor's Response

JOB SHEET 2

Fire Protection

Name: _____   Date: _____

After completing this job sheet, the student should be able to identify the correct fire extinguisher for a particular fire.

Procedures to Prevent and Respond to Fires and Other Hazards

For a fire to burn, it must have three things: fuel, heat, and oxygen. Fuel is anything that can burn, including materials such as wood, paper, cloth, combustible dusts, and even some metals. Fires are divided into five classes: A, B, C, D, and K.

Class A Fires

This class involves common combustible materials such as wood or paper. Class A fire extinguishers often use water to extinguish a fire.

Class B Fires

This class involves fuels such as grease, combustible liquids, or gases. Class B fire extinguishers generally employ carbon dioxide (CO_2).

Class C Fires

This class involves energized electrical equipment. Class C fire extinguishers usually use a dry powder to smother the fire.

Class D Fires

This class involves burning metal. Class D extinguishers place a powder on top of the burning metal that forms a crust to cut off the oxygen supply to the metal.

Class K Fires

This class involves grease fires.

Fire Extinguishers

Identify the correct type of fire extinguisher for the following types of fires.

Fire	Fire Extinguisher Type
Office paper fire	
Office wood fire	
Computer room fire	
Industrial fire (metal)	
Electrical fire	
Gas fire	
Grease fire	
Plastic fire	

Instructor's Response

Pipe, Valves, and Fittings

OBJECTIVES

After completing this chapter, the student should be able to:

- Order certain pipe, valves, and fittings.

KEY TERMS

Pipe

Valve

Fitting

Elbow

Tee

Reducer

ASTM

ANSI

Diameter

Sleeve

SDR

INTRODUCTION

Regardless of the type of work a plumber is asked to do, a good understanding of the products and materials available is essential to effectively design a plumbing system that meets the client's (property owner) needs and conforms to state and local building codes.

REVIEW QUESTIONS

Short Answer

1. What is ASTM?

2. What is potable water?

3. For what is black steel pipe used?

4. How is pipe categorized?

True/False

5. True or False: Standard Dimension Ratio is the ratio of the inside pipe diameter to the wall thickness.

6. True or False: Hard copper pipe is only used in the HVAC industry.

7. True or False: A hub is a socket manufactured to receive the end of another pipe or fitting and is also referred to as a bell.

8. True or False: Valves are used to isolate and regulate portions of a piping system or the entire system.

9. True or False: An offset fitting is typically used to keep the end of one pipe from lining up to another.

10. True or False: A cross is a four-way fitting used to connect two branches to a main pipe.

JOB SHEET **1**

Pipe Identification

Name: _____   Date: _____

After completing this job sheet, the student should be able to identify the correct piping material for a particular task.

Match the pipe material to the application.

_____ **a.** Copper

_____ **b.** Cast iron

_____ **c.** Black steel

_____ **d.** Galvanized steel

_____ **e.** PVC

_____ **f.** ABS

_____ **g.** Polyethylene

_____ **h.** PEX

_____ **i.** CPVC

1. Gas pipe that is connected to a central heating unit

2. Piping that connects a submergible potable pump to a storage tank

3. Primarily used for Drain, Waste, and Vent application (black in color)

4. Yellowish pipe that is used for potable hot water

5. Sold in hard and soft versions and used in most plumbing applications. Also, used by the HVAC industry

6. Used primarily for commercial DWV and storm drainage systems

7. White flexible plastic-type tubing used for potable water and heating systems

8. Black steel pipe lined with a protective coating; this coating allows the pipe to be used for potable water applications

9. Used for cold water applications and DWV

Instructor's Response

JOB SHEET 2

Ordering a Plumbing Tee

Name: **Date:**

After completing this job sheet, the student should be able to order a plumbing tee.

For each tee, correctly provide the ordering information.

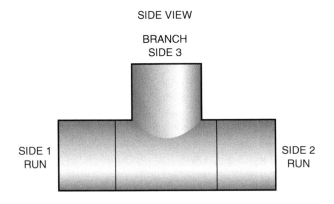

SIDE VIEW

BRANCH
SIDE 3

SIDE 1
RUN

SIDE 2
RUN

Side 1	Side 2	Side 3	Order As
1/2"	1/2"	1/2"	
1/2"	1/2"	3/4"	
3/4"	3/4"	3/4"	
3/4"	3/4"	1/2"	
3/4"	1/2"	1/2"	
3/4"	1/2"	3/4"	
3/4"	3/4"	1"	
1"	1"	1"	
1"	1"	3/4"	
1"	1"	1/2"	
1"	3/4"	3/4"	
1"	3/4"	1/2"	
1"	1/2"	1/2"	

Instructor's Response: Turn page

Instructor's Response

JOB SHEET 3

Valve Identification

Name: _____   Date: _____

After completing this job sheet, the student should be able to correctly identify various valves used in plumbing.

For each valve give a brief description of what that valve is used for.

Type	Use
Pressure-reducing valve	
Check valve	
Vacuum breaker	
Vacuum relief valve	
Relief valve	
Reduced pressure zone valve	
Double check valve assembly	

Instructor's Response

JOB SHEET 4

Fitting Identification

Name: _____   **Date:** _____

After completing this job sheet, the student should be able to correctly identify various fittings used in plumbing.

Write the name of each fitting under the corresponding illustration.

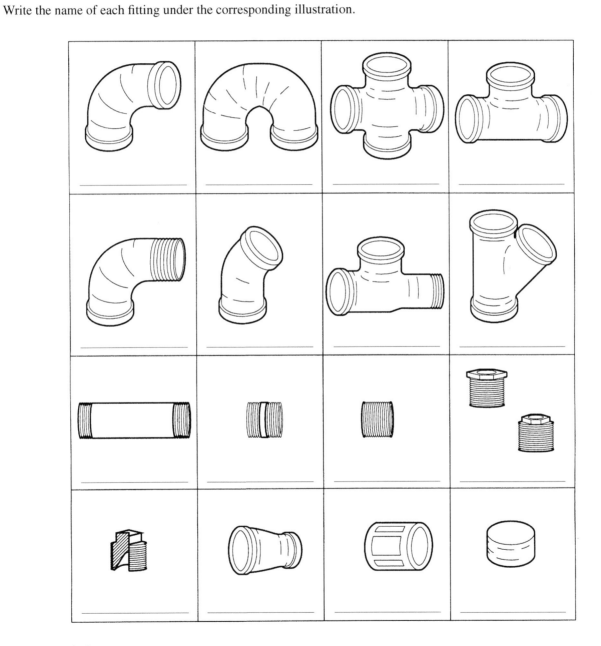

Instructor's Response

Blueprint Reading and Drafting

OBJECTIVES

After completing this chapter, the student should be able to:

- Identify the basic blueprint symbols and abbreviations.
- Interpret a riser diagram.
- Create plumbing drawings and diagrams.

KEY TERMS

Triangle

Riser

Orthographic

View

Diagram

Symbol

Line type

INTRODUCTION

In order to become an effective plumber, it is very important that the student has a good understanding of blueprint reading and basic drawing skills. A plumber is typically not responsible for creating plumbing drawings but is required to follow plumbing drawings for bidding, installation, and, at times, maintenance.

REVIEW QUESTIONS

Short Answer

1. For what is a take-off sheet used?

2. How does an orthographic drawing differ from an isometric drawing?

3. What is the purpose of dimensioning?

4. What is a riser diagram?

5. What are the two common triangles used in drafting?

6. Describe the use of each of the following:

   a. Straightedge

   b. Tee square

   c. Compass

7. In the space provided below, draw a typical drain line.

8. In the space provided below, draw a typical vent.

9. In the space provided below, draw a typical gas line.

10. Complete the following table.

Abbreviation	Stands for
ABS	
ADA	
BS	
DWV	
DH	
FU	
02	
PSI	
PRV	
UNO	

JOB SHEET 1

Symbol Identification

Name: _____   Date: _____

After completing this job sheet, the student should be able to identify some of the symbols typically used in plumbing drawings.

Write in the missing information for the following fittings and valves.

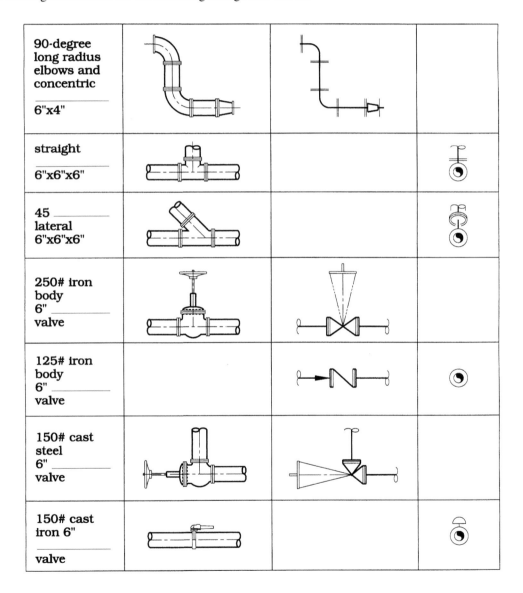

90-degree long radius elbows and concentric 6"x4"			
straight 6"x6"x6"			
45 _____ lateral 6"x6"x6"			
250# iron body 6" valve			
125# iron body 6" _____ valve			
150# cast steel 6" _____ valve			
150# cast iron 6" valve			

Instructor's Response

JOB SHEET 2

Creating Plumbing Drawings

Name: _____   Date: _____

After completing this job sheet, the student should be able to create a basic plumbing drawing.

Refer to the following double-line piping drawing, and create a single-line version.

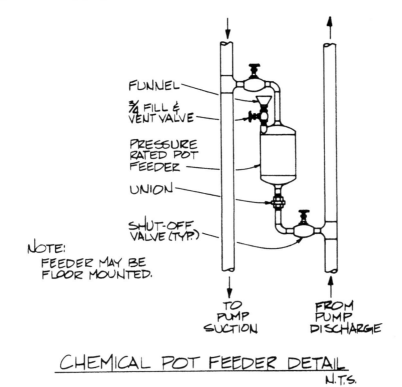

FUNNEL

¾ FILL &
VENT VALVE

PRESSURE
RATED POT
FEEDER

UNION

SHUT-OFF
VALVE (TYP.)

NOTE:
    FEEDER MAY BE
    FLOOR MOUNTED.

TO
PUMP
SUCTION

FROM
PUMP
DISCHARGE

CHEMICAL POT FEEDER DETAIL
N.T.S.

Instructor's Response: Turn page

Instructor's Response

JOB SHEET 3

Working with Plumbing Drawings

Name: _____   **Date:** _____

After completing this job sheet, the student should be able to interpret a plumbing drawing.

Fill in the blanks in the following drawings. Please note that some of the information cannot be calculated. In this case, give a brief description about why it cannot be calculated.

1/16" gasket is used for dimensioning.

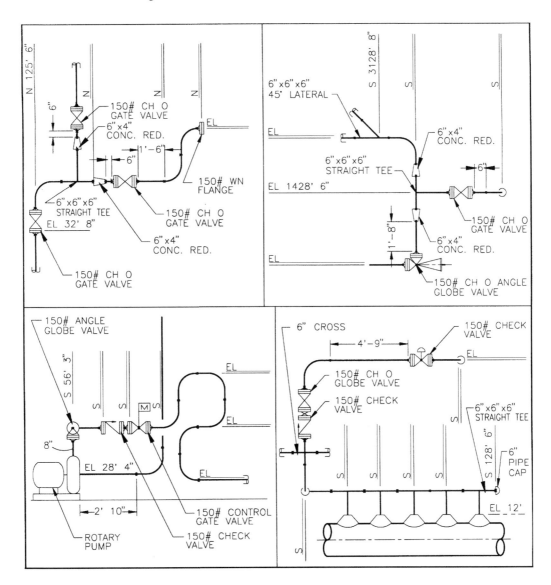

Instructor's Response

Trade Math and Formulas

OBJECTIVES

After completing this chapter, the student should be able to:

- Understand cylindrical capacity formulas.
- Perform basic metric conversions.
- Convert decimal and fractions.
- Work with pipe offset formulas.

KEY TERMS

English Rule

Metric

Length

Width

Depth

Decimal

Fraction

INTRODUCTION

As the world becomes more and more technical and building trades evolve, the need for basic math skills reduces, or so it may seem. In reality, the need to perform basic math has not decreased with the introduction of computer and calculators but instead has remained the same and, in many cases, it has increased.

REVIEW QUESTIONS

Short Answer

1. List the two systems of measurement.

2. What is a piping offset?

3. What is the formula for calculating the area of a circle?

4. What is the formula for calculating a volume of a cube?

5. Define fitting knockout.

JOB SHEET 1

Working with Whole Numbers

Name: _____   Date: _____

After completing this job sheet, the student should be able to add, subtract, multiply, and divide common whole numbers.

Add the following whole numbers. (Circle the correct answer.)

1. 1604
    4488
    + 243

    a. 6,337   **c.** 6,333
    b. 6,335   **d.** 6,213

2. 3551
    1460
    + 1732

    a. 6,743   **c.** 6,735
    b. 20,602   **d.** 5,203

3. 1396
    1925
    + 3527

    a. 6,851   **c.** 6,848
    b. 13,901   **d.** 4,496

4. 1140
    2694
    + 1220

    a. 5,054   **c.** 5,044
    b. 14,810   **d.** 3,969

5. 2782
    5361
    + 5217

    a. 13,362   **c.** 13,358
    b. 13,360   **d.** 10,751

6.   322
    2333
    + 707

    a. 3,362   **c.** 3,354
    b. 9,018   **d.** 2,734

7.   4168
      1204
+ 2501

   a. 7,876   **c.** 7,873
   b. 12,876   **d.** 6,205

8.   3911
      3567
+   194

   a. 7,672   **c.** 7,663
   b. 9,226   **d.** 7,499

9.   2740
      1828
+ 1379

   a. 5,949   **c.** 5,945
   b. 5,947   **d.** 5,257

10.   4687
       393
+ 2868

   a. 7,948   **c.** 7,940
   b. 30,897   **d.** 5,399

Subtract the following whole numbers. (Circle the correct answer.)

11. 2035 − 1202

   a. 836   **c.** 833
   b. 2499   **d.** 277

12. 1034 − 945

   a. 93   **c.** 85
   b. 356   **d.** 89

13. 1955 − 1635

   a. 320   **c.** 311
   b. 2880   **d.** 35

14. 3878 − 1226

   a. 2654   **c.** 2650
   b. 2652   **d.** 1326

15. 4534 − 2444

   a. 2090   **c.** 2081
   b. 18810   **d.** 232

Multiply the following whole numbers. (Circle the correct answer.)

16.   1755
   $\times$ 3565

   a. 6,256,575   **c.** 6,254,941
   b. 33,837   **d.** 694,994

17.   2051
   $\times$ 4270

   a. 8,756,495   **c.** 8,756,491
   b. 19,130   **d.** 8,757,770

18.   2396
   $\times$ 49

   a. 116,392   **c.** 117,404
   b. 2,541   **d.** 38,797

19.   1154
   $\times$ 961

   a. 4,038   **c.** 1,108,994
   b. 1,109,057   **d.** 369,685

20.   5808
   $\times$ 4999

   a. 9,677,519   **c.** 29,034,192
   b. 20,805   **d.** 2,939,412

Divide the following whole numbers. (Write in the correct answer.)

21. 931 ÷ 19

22. 1150 ÷ 46

23. 306 ÷ 34

24. 2597 ÷ 49

25. 858 ÷ 39

26. 00 ÷ 00

Complete the following:

27. $362 + 1{,}491 + 73 + 29{,}248 =$ _____

28. $4{,}793 - 404 =$ _____

29. $189 \div 9 =$ _____

30. $13{,}328 \div 238 =$ _____

Instructor's Response

JOB SHEET 2

Working with Fractions

Name: _____   Date: _____

After completing this job sheet, the student should be able to add, subtract, multiply, and divide fractions.

Add the following common fractions. (Write in the correct answer.)

1. $^6/_{13}$
      $+ \ ^8/_9$
      _____

2. $^{19}/_9$
      $+ \ ^{14}/_{11}$
      _____

3. $^{18}/_{17}$
      $+ \ ^6/_{14}$
      _____

4. $^4/_1$
      $+ \ ^{19}/_4$
      _____

5. $^{10}/_{16}$
      $+ \ ^4/_{19}$
      _____

Subtract the following common fractions. (Write in the correct answer.)

6. $^3/_1$
      $- \ ^{17}/_5$
      _____

7. $^1/_{17}$
      $- \ ^3/_7$
      _____

8. $^{20}/_{20}$
      $- \ ^{15}/_{10}$
      _____

9. $^{20}/_9$
      $-\,^{14}/_{19}$
      _____

10. $^{12}/_{19}$
       $-\,^{13}/_{15}$
       _____

Multiply the following common fractions. (Write in the correct answer.)

11. $^2/_{20}$
       $\times\;^{16}/_4$
       _____

12. $^{20}/_{16}$
       $\times\;^2/_6$
       _____

13. $^{19}/_{19}$
       $\times\;^{14}/_9$
       _____

14. $^{19}/_8$
       $\times\,^{13}/_{18}$
       _____

15. $^{11}/_{18}$
       $\times\,^{12}/_{14}$
       _____

Divide the following common fractions. (Write in the correct answer.)

16. $^4/_3 \div\, ^{18}/_6$

17. $^3/_{19} \div\, ^4/_8$

18. $^2/_2 \div {}^{17}/_{11}$

19. $^2/_{10} \div {}^{15}/_1$

20. $^{14}/_{20} \div {}^{15}/_{17}$

Instructor's Response

JOB SHEET 3

Unit Conversions

Name: _____   Date: _____

After completing this job sheet, the student should be able to convert from metric to English Rule and English Rule to metric.

1. Convert 40 square meters to square millimeters.

2. Convert 10 square feet to square inches.

3. Convert 104 square meters to square inches.

4. Convert 20 square meters to square millimeters.

5. Convert 92 square feet to square inches.

6. Convert 143 square meters to square inches.

7. Convert 11 square meters to square millimeters.

8. Convert 4 square feet to square inches.

9. Convert 200 square meters to square inches.

10. Convert 12 square meters to square millimeters.

Instructor's Response

JOB SHEET 4

Area and Volume

Name: _____   Date: _____

After completing this job sheet, you should be able to calculate the volume of a sphere, cylinder, cube, and so on, as well as the areas of various geometric shapes.

1. The circumference of a circle with a 5.62 centimeter diameter is _____. (Round to three significant digits.)

2. The radius of a circle that has a circumference of 48' 9" is _____. (Round to the nearest inch.)

3. The length of a 67.37° arc on a circle with a 2.765-inch radius is _____. (Round to four significant digits.)

4. The diameter of a circle which has an arc length of 36.90 centimeters and a central angle of 128°18' is _____. (Round to four significant digits.)

5. The area of a rectangle 18.0 inches long and 5.60 inches wide is _____.

6. The area of a rectangle is 146 square feet and the length is 18.30 feet. The width is _____.

7. The volume of a prism with a height 4.80 inches and a base area 150.0 square inches is _____.

8. Compute the volume of a right circular cylinder that has a base area of 246 square centimeters and a height of 8.20 centimeters.

9. A cylindrical hot water tank is 1.50 meters high and has a diameter of 0.400 meter.

    a. Compute the volume of the tank.

    b. How many liters of water are held in the tank when full?

Instructor's Response

SECTION 2

Fixtures and Faucets

6 Fixture Types

OBJECTIVES

After completing this chapter, the student should be able to:

- Identify common plumbing fixtures.

- Know basic fixture designs.

- Understand the Americans with Disabilities Act.

KEY TERMS

ADA

Carrier

China

EWC

Grid strainer

Pop-up

Vortex

Fixture

INTRODUCTION

Calculating and installing pipe is only a small portion of a plumber's job; the plumber must also be concerned with plumbing fixtures. A plumbing fixture is defined as *a device which is part of a system to deliver and drain away water, but which is also configured to enable a particular use.*

RESTROOMS AND BATHING FACILITIES

The ADA requires at least one restroom to be accessible to people with disabilities for each sex, per accessible floor.

In some cases a unisex restroom can satisfy this requirement. The door to the restrooms must meet all door requirements shown in Figure 6–1, including the distance between doors at the vestibule. All of the fixtures intended for people with disabilities shall be on an accessible route. This means that there must be adequate access to the fixtures, and the restroom must have room for a wheelchair to maneuver. Figure 6–2 shows the minimum acceptable maneuvering space. A preferred maneuvering configuration is shown in Figure 6–3.

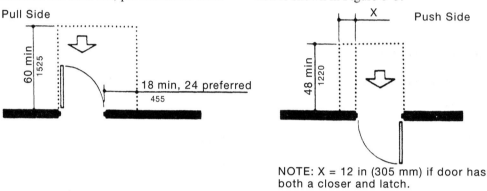

(a)
Front Approaches – Swinging Door

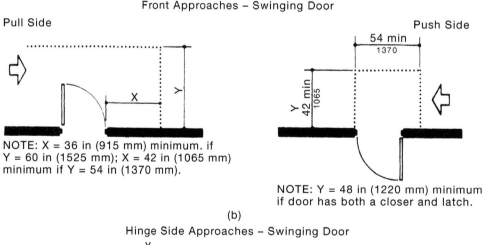

(b)
Hinge Side Approaches – Swinging Door

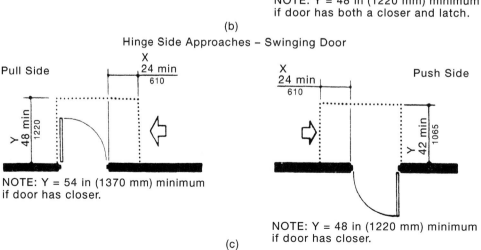

(c)
Latch Side Approaches – Swinging Door

NOTE: All doors in alcoves shall comply with the clearances for front approaches

Maneuvering Clearances at Doors

Figure 6–1

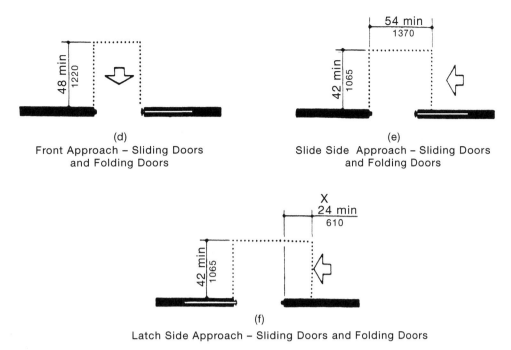

(d)
Front Approach – Sliding Doors
and Folding Doors

(e)
Slide Side Approach – Sliding Doors
and Folding Doors

(f)
Latch Side Approach – Sliding Doors and Folding Doors

NOTE: All doors in alcoves shall comply with the clearances for front approaches

Maneuvering Clearances at Doors

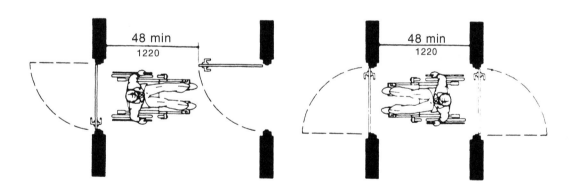

Two Hinged Doors in Series

Figure 6–1   *(Continued)*

Toilet Stalls

At least one toilet stall in each restroom shall meet the requirements shown at A in Figure 6–4. These stalls may be reversed to allow for either a left- or right-handed approach. Alternate stall placement (at B in Figure 6–4) is not allowed in new construction. If a restroom contains six or more stalls, in addition to the standard accessible stall, one of these stalls must be at least 36" (915 mm) wide with an outward swinging door. The front and one side partition of the stall must be at least 9" (230 mm) above the floor unless the stall is longer than 60" (1525 mm). A single-person restroom with a toilet and a sink without a stall shall meet the requirements shown in Figure 6–5. The drafter should detail door swings to stalls to maintain required clearances.

The toilet or water closet shall be 17" to 19" (430 to 485 mm) to the top of the toilet seat. Stalls for wall-mounted toilets can be 3" (75 mm) shorter than those for floor-mounted toilets. Grab bars, shown in Figure 6–4,

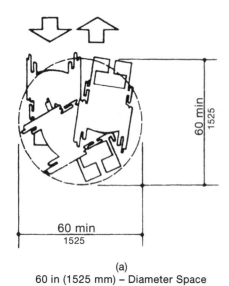

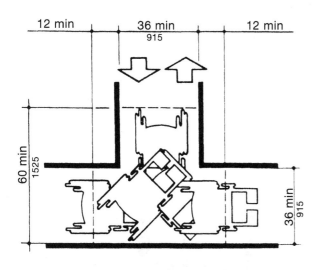

(a)
60 in (1525 mm) – Diameter Space

(b)
T-Shaped Space for 180° Turns

Figure 6–2

shall meet the requirements illustrated at E in Figure 6–6. The toilet paper dispenser shall be 19" to 36" (485 to 915 mm) above the floor.

Urinals

If urinals are provided, at least one shall meet the following requirements: It shall be stall-type or wall-hung with an elongated rim at a maximum of 17" (430 mm) above the floor. A clear floor space of 30–48" (815–1220 mm) shall be provided in front of the urinals.

This clear space can overlap the accessible route. Partitions that do not project beyond the rim can be installed with 29" (740 mm) minimum clearance.

Lavatories and Mirrors

If lavatories and mirrors are provided, at least one shall meet the following requirements: Lavatories and mirrors shall conform to the requirements shown in Figure 6–7. The clear space can overlap the accessible route. Hot water and drain pipes under lavatories shall be insulated or otherwise configured to protect against contact. There shall be no sharp or abrasive surfaces under the lavatories. Faucets shall be operable with one hand and without requiring a tight grip, tight pinching, or twisting of the wrist. They shall be operable with a maximum force of 5 lbs (2.3 kg). Lever-operated, push-type and electrically operated faucets are examples of acceptable designs. If self-closing valves are used, the faucet shall remain open for at least 10 seconds.

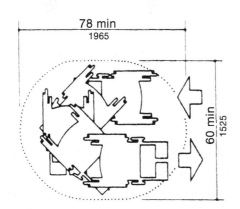

Space Needed for Smooth U-Turn in a Wheelchair

Figure 6–3

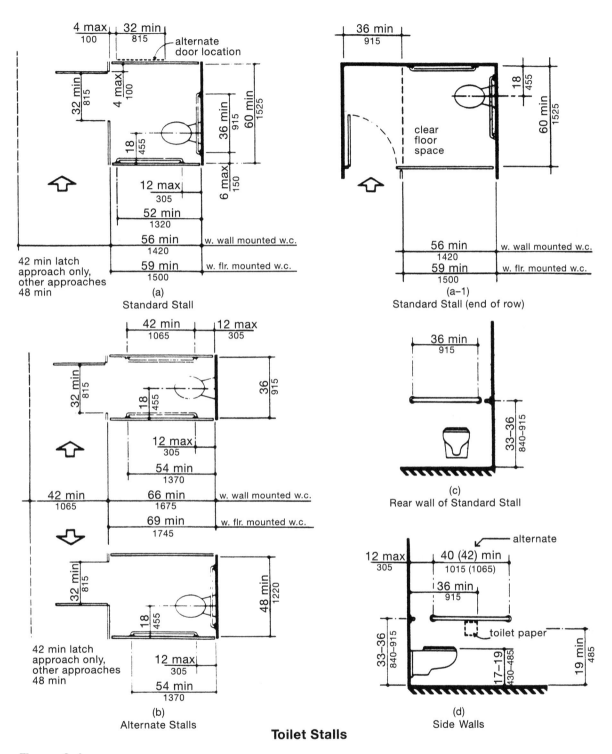

Toilet Stalls

Figure 6–4

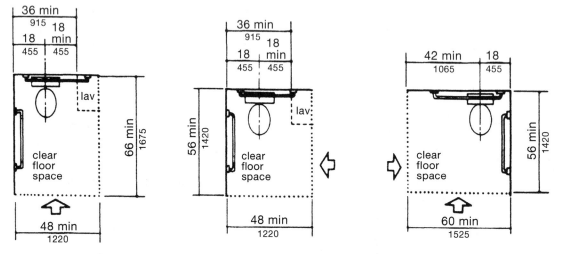

Clear Floor Space at Water Closets

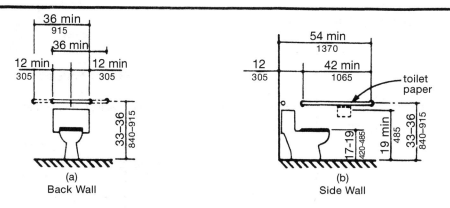

Grab Bars at Water Closets

Figure 6–5

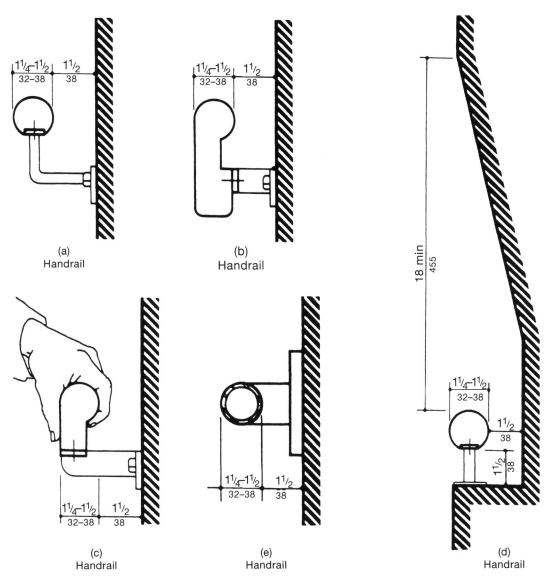

Figure 6–6

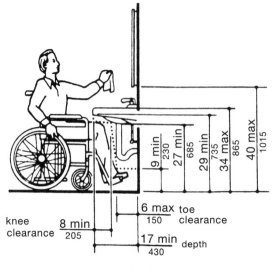

Lavatory Clearances

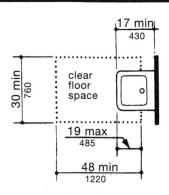

Clear Floor Space at Lavatories

Figure 6–7

REVIEW QUESTIONS

True/False

1. True or False: All plumbing fixtures are manufactured to strict federal codes and therefore when selecting a plumbing fixture the plumber is only concerned with the price and the fixtures function.

2. True or False: The water supply and DWV servicing the plumbing fixture is typically installed during the rough-in phase of a project.

3. True or False: A sink will typically have a 1 1/4" drain while a lavatory will have a 1" drain.

4. True or False: A water connection to a urinal is typically completed by a part known as a nipple.

5. True or False: Most common style bathtubs are 60" in length and 30" wide.

6. True or False: The American with Disabilities Act requires full access to a public fixture.

Matching

7. Match the following fixtures with their descriptions.

    a. Toilet

    b. Lavatory

    c. Urinal

    d. Bathtub

    e. Shower

    f. Sink

_____ is also known as a basin.

_____ are categorized differently from basins because they are capable of holding more gallons of water, which results in more water draining into a piping system.

_____ is referred to as a tub in the plumbing industry and are available in a various types, colors and designs.

_____ are typically sold as one-piece units and multiple piece units.

_____ are rarely installed in residential applications and are hung from the wall.

_____ are also known as a water closet and are available in many styles and colors.

Completion

8. Specialty fixtures are typically used in _____, _____, and/or _____.

Short Answer

9. How are drinking fountains ordered?

10. What is the minimum shower size allowed by nonhandicap codes?

JOB SHEET **1**

ADA Requirements for a Toilet

Name: _____   Date: _____

After completing this job sheet the student should be able to understand the ADA requirements for a toilet.

Currently you are working for a plumbing contractor and you are asked to work on the following project:

A local retail department store is expanding and as a result they are adding another group of male and female restrooms. The new men's restroom will have three urinals and five toilets. The new women's restroom will have five toilets. All fixtures used in the facilities are to be white.

1. What are the ADA requirements for the new men's and women's restroom?

2. Create an order form for the toilets.

Instructor's Response

JOB SHEET 2

ADA Requirements for Lavatory

Name: _____   Date: _____

After completing this job sheet, the student should be able to understand the ADA requirements for a lavatory.

Refer to the information regarding the facility in job sheet 1. The men's restroom contains three white oval countertop lavatories, each having three holes with an 8" spread. The women's restroom will also have three white oval countertop lavatories, each having three holes with an 8" spread.

1. Determine the ADA requirements for the lavatories for both the men's and women's restrooms.

2. Create an order form for the lavatories.

Instructor's Response

7 Faucets and Drain Assemblies

OBJECTIVES

After completing this chapter, the student should be able to:

- Identify common faucets and drain assemblies.
- Identify basic ordering criteria for faucets and drain assemblies.

KEY TERMS

Aerator

Backflow

Basket strainer

P.O.

P-trap

Flanged tailpiece

INTRODUCTION

When selecting a faucet to install, there are several things that a plumber must consider and often the final selection will be made by the facility's owner.

- Selecting which manufacturer's product to install is based on cost, quality, and preferred faucet design.

- Most master bathrooms and guest bathrooms have more expensive finishes than other bathrooms in a house.

- All faucet finishes dictate the finishes used for drain assemblies and bathroom accessories to create a color theme.

- Faucet accessories are available to create various themes.

REVIEW QUESTIONS

Short Answer

1. What is the purpose of the drain assembly?

2. What is the purpose of the air gap?

3. What is the main requirement for selecting a kitchen sink faucet?

Completion

4. Pop-ups are abbreviated as _____ and are most common in _____ and light _____ installations.

True/False

5. True or False: A typical three-hole lavatory sink has a spread of 6" and is often referred to as a center set. If false explain why.

6. True or False: The kitchen sink drain assembly does not require a basket strainer. If false explain why.

7. True or False: All lavatory sinks have an overflow slot. If false explain why.

8. True or False: All shower drains require a safety pan to be installed. If false explain why.

9. True or False: A laundry sink will either have a drain connection manufactured as part of the sink, or will use a jr. basket strainer. If false explain why.

10. True or False: If the hygiene spray is below the flood level rim of the bidet than a vacuum breaker or other backflow prevention device must be installed. If false explain why.

JOB SHEET 1

Faucet and Drain Questions

Name: _____   Date: _____

After completing this job sheet, the student should be able to identify the common faucet and drain assemblies.

Completion

1. _____ is a flange installed to conceal pipe penetrations through a wall, floor, or ceiling.

2. _____ is the color or polish of a faucet, drain assembly, or other fixture trim item.

3. A tub faucet is mounted either on the _____ or _____.

4. _____ faucets are installed with large capacity tubs, such as a garden tub or whirlpool.

Matching

Match the following terms with the corresponding definitions.

_____ **a.** Escutcheon

_____ **b.** Finish

_____ **c.** Port opening

_____ **d.** Pop-up

5. An opening in a fixture, such as a drain or overflow hole that receives drain assemblies, to connect the fixture to the drain system

6. Drain assembly for lavatory sinks and bidets and abbreviated as P.O.

7. The color or polish of a faucet, drain assembly, or other fixture trim item

8. Flange installed around a pipe to conceal pipe penetrations through a wall, floor, or ceiling

Instructor's Response

JOB SHEET 2

Faucet Identification

Name: _____   **Date:** _____

After completing this job sheet, the student should be able to identify the various parts of a faucet.

Write in labels to correspond with arrows in the following illustration.

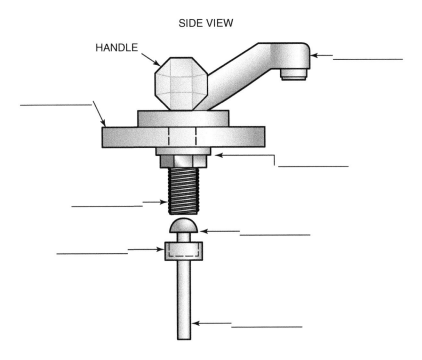

Instructor's Response

Bidet Drain Assembly

Name: _____   **Date:** _____

After completing this job sheet, the student should be able to identify the various parts of a bidet drain assembly.

Write in labels to correspond with arrows in the following illustration.

SIDE VIEW

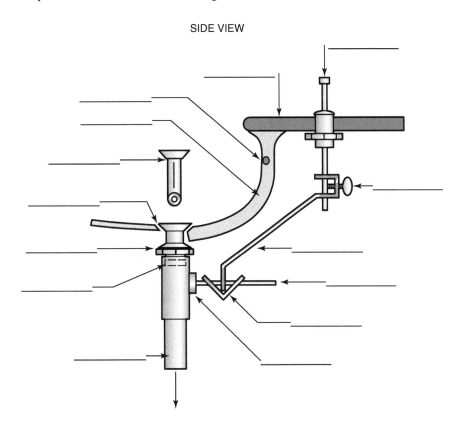

Instructor's Response

Fixture Installations

OBJECTIVES

After completing this chapter, the student should be able to:

- Understand basic steps in installing common fixtures.
- Understand fixtures connection variations.
- Appreciate the importance of accurate rough-in piping.

KEY TERMS

Ferrule

Slip joint

Spud

Stub-out

Stub-up

Tailpiece

INTRODUCTION

Once the rough-in phase of the construction has been completed and the facility has been enclosed, the trim-out process usually begins. The trim-out phase of the project is when the actual fixtures are installed and tested. In some cases, however, it is necessary to install fixtures during the rough-in phase. In this case, it is recommended that the plumbing technician cover the fixture with cardboard to protect the finish. Unless you are employed by a large plumbing contractor chances are you will be doing both rough-in and trim work. Therefore, it is essential that you become proficient in both phases of a plumbing project.

REVIEW QUESTIONS

Completion

1. _____ is one aspect of a compression joint. It is a sleeve placed over a tube and compressed around the tube to form a seal when tightened with a compatible nut.

2. _____ is a method of connecting under sink drain piping that utilizes special washers and tightening nuts.

3. If a commercial toilet is served with a flush valve, the water supply rough-in pipe will be _____ inches to the left or right of the center of the toilet.

4. An _____ known as a stop is installed onto the stub-out or stub-up pipe.

5. The drain from a dishwasher connects either to a specially designed _____ tailpiece or to dedicated drain inlet of a garbage disposal unit.

6. If no size is specified then all drainage piping and p-traps associated with a lavatory should be a minimum of _____ .

True/False

7. True or False: Some installations use a stub-up in which the water pipe protrudes vertically from the floor under the left side of the toilet tank. If false explain why.

8. True of False: Most large commercial projects use compression stops while most residential projects use soldered stops. If false explain why.

9. True or False: Commercial kitchens use a variety of sink styles that are mostly stainless steel and free-standing. If false explain why.

Short Answer

10. What is a dishwasher elbow?

JOB SHEET **1**

Toilet Installation

Name: _____   **Date:** _____

After completing this job sheet, the student should be able to identify components used in the installation of a toilet.

Write in labels to correspond with arrows in the following illustrations.

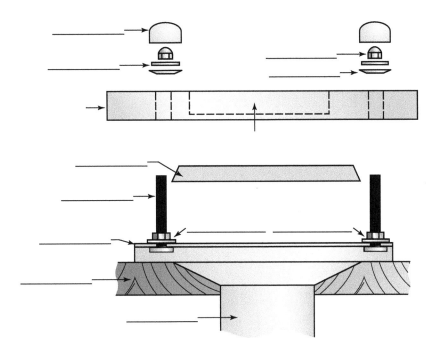

Instructor's Response

JOB SHEET 2

Lavatory Installation

Name: _____ Date: _____

After completing this job sheet, the student should be able to identify components used in the installation of a lavatory.

Write in labels to correspond with arrows in the following illustration.

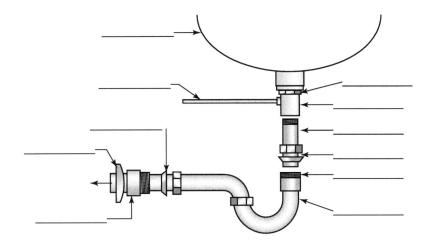

Instructor's Response

JOB SHEET 3

Bidet Installation

Name: _____   Date: _____

After completing this job sheet, the student should be able to identify components used in the installation of a bidet.

Write in labels to correspond with arrows in the following illustration.

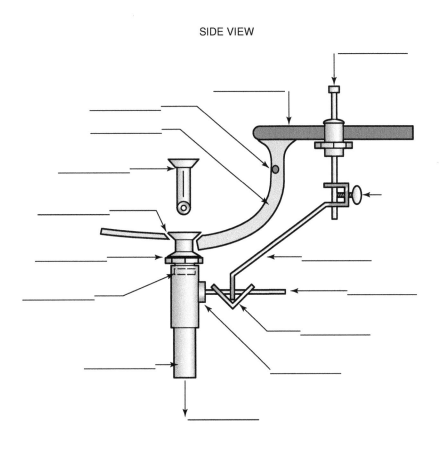

SIDE VIEW

Instructor's Response

SECTION 3

Water Supply Systems

Introduction to Water Supply

OBJECTIVES

After completing this chapter, the student should be able to:

- Understand sources of potable water.
- Recognize the differences between private and public water systems.
- Explain and determine water pressure and hydraulic pressure.

KEY TERMS

Cased

Cistern

Drilled wells

Driven wells

Dug wells

Head

Potable

Pressure reducing valve

Water hammer

Water table

INTRODUCTION

Potable water is a primary concern of most plumbing jobs. Without an adequate potable water supply it becomes impossible to maintain many of the modern conveniences and services that are necessary to sustain the quality of life to which our society has grown accustomed. Potable water is used not only for drinking water but for any application that might come in contact with humans.

REVIEW QUESTIONS

Short Answer

1. Describe the water cycle.

2. What is potable water?

True/False

3. True or False: A well system is controlled by a pressure switch, which typically does not allow the water pressure to exceed 60 psi and may not require a pressure-reducing valve by some codes. If false explain why.

4. True or False: The intensity of water hammering depends upon the volume of air being pushed by the water. If false explain why.

5. True or False: Pressure on any part of an enclosed liquid is exerted uniformly in a single direction. If false explain why.

Completion

6. A _____ pump is limited to a theoretical depth of 35 feet.

7. The _____ is a federal law regulating public water systems; it creates guidelines for local municipalities to define what is considered a public water system.

8. _____ is caused by a moving body of water suddenly stopping in a pipe.

9. _____ breaks a vacuum in a piping system by introducing air into the pipe.

10. _____ is a single device in which two spring-loaded check assemblies are incorporated.

JOB SHEET 1

Hydrostatic Pressure

Name: _____   Date: _____

After completing this job sheet, the student should be able to calculate fluid pressure.

Fluid Pressure Formula:

Pressure = Force (Pounds) $\times$ Area (Square Inches)

Note: Pressure will be expressed in Pounds/Square Inch.

Calculate the fluid pressure for the following example.

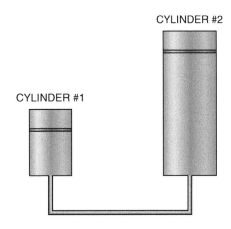

Cylinder #1

Diameter = 1"

Depth of Cylinder = 3"

Cylinder #2

Diameter = 2"

Depth of Cylinder = 12"

Force applied to cylinder = 200 Pounds

Hint: (Force for Cylinder #1)/(Area Cylinder #1) = (Force for Cylinder #2)/(Area for Cylinder #2)

Instructor's Response

JOB SHEET 2

Fluid Flow Rate

Name: _____   Date: _____

After completing this job sheet, the student should be able to calculate fluid flow rate.

Fluid Flow Rate Formula:

Flow Rate = Volume (Gallons) $\times$ Time (Minutes)

Note: Flow rate will be expressed in gallons/minute

Calculate the flow rate for the following:

1. 2" pipe carrying 5 gallon of water in 15 minutes with 40 PSI of pressure.

2. 1" pipe carrying 15 gallon of water in 12 minutes with 50 PSI of pressure.

3. 1" pipe carrying 9 gallon of water in 9 minutes with 60 PSI of pressure.

4. 3" pipe carrying 35 gallon of water in 20 minutes with 50 PSI of pressure.

5. 2" pipe carrying 7 gallon of water in 2 minutes with 65 PSI of pressure.

Instructor's Response

JOB SHEET 3

Water Meter Installation for Cold Weather Climates

Name: _____   Date: _____

After completing this job sheet, the student should be able to identify the components of a cold climate water meter installation.

Write in labels to correspond with arrows in the following illustration.

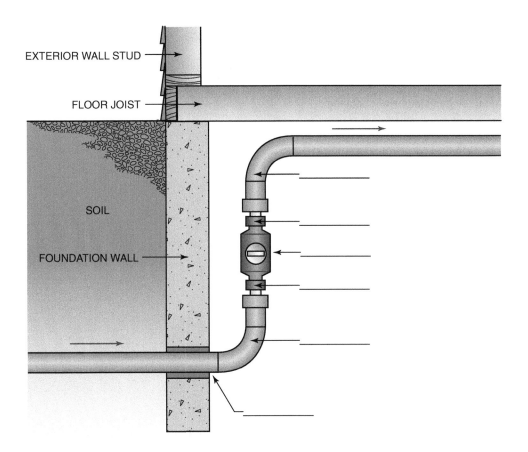

Instructor's Response

OBJECTIVES

After completing this chapter, the student should be able to:

- Understand water rough-in criteria for basic fixtures.
- Identify the various types of hangers and explain their uses.
- Comprehend the basic steps of installing common types of water piping systems.

KEY TERMS

Capillary action

Circulate

Hanger

Flux

Metal filler

INTRODUCTION

The hot and cold water piping in a facility is known as the water distribution system. The installation of water distribution systems is different for commercial and residential in the type of equipment installed as well as the size of pipe used. In addition to this, the codes that regulate the installation of plumbing systems differ.

REVIEW QUESTIONS

Completion

1. _____ is a paste used to promote the flow of solder into a fitting.

2. _____ is the movement of water in a hot water piping system by a pump that returns water to a water heater.

3. _____ is a channel iron support used for numerous purposes, primarily for supporting pipes in a manner known as a trapeze hanger.

True/False

4. True or False: Flexible tubing such as PEX usually requires support to be installed a maximum of 32". If false explain why.

5. True or False: The minimum pipe size allowed by code is always the actual size installed by a plumber. If false explain why.

6. True or False: The fixture rough-in is determined by the specific fixtures being installed. If false explain why.

7. True or False: Commercial toilet rough-in requires a more exact placement than residential toilet installation. If false explain why.

8. True or False: A kitchen sink can be served by either a stub-out or a stub-up. If false explain why.

9. True or False: A lavatory rough-in is always installed with a stub-out. If false explain why.

10. True of False: PVC is not allowed by code to be installed for a water distribution system. If false explain why.

JOB SHEET 1

Soldering and Brazing

Name: _____   Date: _____

After completing this job sheet, the student should be able to make connections with copper tubing using both low-temperature solder and high-temperature braze material.

Safety Precautions: You will be working with acetylene gas and heat. Your instructor must also provide you with instruction in soldering and brazing before you begin this exercise. Wear goggles and light gloves while soldering or brazing.

Procedure:

1. Clean the ends of the tubing with the sand cloth.

2. Clean the inside of the fittings using sand cloth or brushes.

3. Apply flux to the outside of the pipe and to the inside of the fittings.

4. Fit the assembly together using the correct flux.

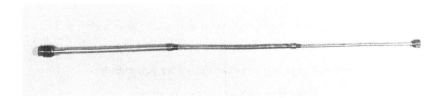

5. Place the assembly in a vertical position using the vise. This will give you practice in soldering both up and down in the vertical position.

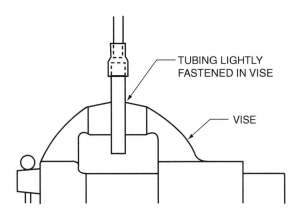

TUBING LIGHTLY
FASTENED IN VISE

VISE

6. Put on your goggles. Follow procedures outlined by your instructor for setting up and using acetylene gas equipment. Start with the top joint at the top connection, then solder the joint underneath. After completing this, solder the top joint of the second connection, then solder underneath. Be sure not to overheat the connections.

   CAUTION: If the tubing becomes red or discolors badly, the connection is too hot.

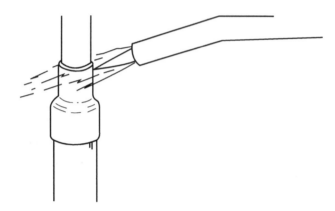

7. Let the assembly cool before you disturb it or the connections may loosen.

8. Cut the reducing connectors out of the assembly and re-clean the tubing ends.

9. Clean the second set of reducing couplings.

10. Assemble the pipes and connectors again using flux, and hold them upright in the vise.

11. Put on your goggles, solder the top connections, then solder the bottom connections as in the previous exercise. Only use high-temperature brazing material.

    CAUTION: the pipe and fittings will have to be cherry red to melt the filler metal, so be careful.

12. After the fittings have cooled, leak-test them as before. Wear goggles and gloves while transferring refrigerant.

13. You can use a hacksaw to cut into the fittings and examine the solder penetration.

14. Upon completion of the leak test, cut off the flare nuts and scrap the tubing.

Maintenance of Workstation and Tools: Clean any flux off the tools or fittings and the work bench. Turn off the torch, turn off the tank, and bleed the acetylene from the torch hose. Place all extra fittings and tools in their proper places.

Answer the following questions:

1. What is the approximate melting temperature of 50/50 solder?

2. What is the approximate melting temperature of 45% silver solder?

3. What is the approximate melting temperature of 15% silver solder?

4. Should the flame or the tube melt the solder in the soldering process?

5. What is the metal content of 50/50 solder?

6. What is the metal content of 95/5 solder?

7. Which solder is stronger, 95/5 or 15% silver solder? Why?

8. Which solder would be the best choice for the discharge line, 95/5 or silver solder? Why?

9. What pulls the solder into the connection when it is heated? Explain this process.

10. Why is a special sand cloth used on hermetic compressor systems?

Instructor's Response

SECTION 4

Water Heating

OBJECTIVES

After completing this chapter, the student should be able to:

- Understand basic water heating theory.
- Interpret British thermal units.

KEY TERMS

Atmosphere

Boiling point

British thermal unit

Normal atmospheric pressure

Normal boiling point

Recovery rate

Temperature rise

INTRODUCTION

To better understand how to select a water heater you must first have a good understanding of the principles that govern heat transfer and the properties of water. Although heat transfer is the core of the HVAC industry, a basic understanding is still essential if you are planning to work with water heaters and hydronic.

REVIEW QUESTIONS

Completion

1. _____ is the amount of heat required to raise one pound of water one degree Fahrenheit.

2. _____ is the temperature at which a liquid changes to a gas (vapor).

3. _____ is the difference in incoming cold water and outgoing hot water of a water heater.

4. _____ is the unit of pressure roughly equal to the average atmospheric pressure at sea level on the earth.

5. _____ is pressure exerted onto earth and is the same as one atmosphere or 14.7 psi.

6. _____ is the performance of a water heater based on how many gallons per hour a water heater can produce.

7. _____ is the temperature at which liquid boils when the external pressure is one atmosphere.

JOB SHEET 1

Calculating BTUs

Name: _____   Date: _____

After completing this job sheet, the student should be able to calculate btus at different atmospheric pressures.

Circle or write in the correct answer.

1. The standard atmospheric conditions for water to boil at 212°F are:

   a. Atmospheric pressure: 15.7 psi, temperature: 70°F

   b. Atmospheric pressure: 16.7 psi, temperature: 68°F

   c. Atmospheric pressure: 14.7 psi, temperature: 70°F

   d. Atmospheric pressure: 15.7 psi, temperature: 68°F

2. If water weighs 8.33 pounds per gallon, it will take _____ btus to raise the temperature of 4 gallons of water from 70°F to 72°F.

   a. 66.6   c. 14.7

   b. 10   d. 106

3. The atmosphere will support a column of mercury _____ inches at sea level.

   a. 32.00   c. 14.7

   b. 28.69   d. 29.92

4. Calculate the amount of heat to raise the temperature of 5 pounds of water 75°F at sea level.

5. Calculate the amount of heat to raise the temperature of 100 pounds of water 132°F at sea level.

Instructor's Response

Water Heater Types and Installations

OBJECTIVES

After completing this chapter, the student should be able to:

- Describe the different types of water heaters.
- Understand the basics of installing a water heater.

KEY TERMS

Aerator

Anode rod

Boiler plate

Bottom plate

Bottom-fed

Dip tube

Flue gasses

Heating elements

T&P

Tankless

Top-fed

INTRODUCTION

The basic principles that govern water heaters are the same for commercial applications as well as residential applications; however, the size of the units varies greatly due to the increased demand that is typically associated with commercial applications. For example, because of health requirements restaurants will have a greater demand for hot water than a residence.

REVIEW QUESTIONS

Completion

1. _____ is a label applied to the exterior of a water heater indicating specific information about the water heater.

2. _____ are the deadly fumes from a gas water heater that are evacuated from a building to the exterior through a flue pipe.

3. _____ is the abbreviation for temperature and pressure.

4. The _____ are immersed in the water of an electric water heater to heat the water inside a storage tank.

5. _____ is the internal tube installed by the manufacturer to funnel incoming cold water of a top-fed storage tank to the bottom of a tank.

6. _____ is a removable flow accessory attached to many faucet spouts to create a uniform flow of water.

7. _____ indicates that the cold water connection of a water heater is located on the bottom and the side of a storage tank and does not use a dip tube.

8. _____ is a type of water heater that is other than a storage tank type.

9. _____ indicates the cold water connection to a storage tank is located on the top of the storage tank and uses a dip tube to funnel water to the bottom of a tank.

JOB SHEET 1

Common Storage Tank Drainage Sequence

Name: _____   Date: _____

After completing job sheet, the student should be able to follow the correct sequence for draining a storage tank.

Use the comment column in the following table to list when each step was completed, who complete the step, and/or any other special circumstances that might have arisen during the completion of that step.

Step	Comment
1. Turn off cold water.	
2. Turn off heating source.	
3. Connect garden hose and, if required, connect drain pump.	
4. Route hose to discharge in safe location.	
5. Open T&P value to allow air to enter the piping system.	
6. Open boiler drain and if relevant, energize drain pump until tank is drained.	
7. Proceed with disconnection of water pipes, and gas pipe or electric wiring.	

Instructor's Response

Basic Sequence for Replacing a Water Heater

Name: _____ Date: _____

After completing this job sheet, the student should be able to replace a water heater.

When replacing a water heater, use the same type. For example, if you have an electric water heater, replace it with an electric water heater unless you are willing to run gas lines and exhaust vents. If you are replacing a gas water heater, replace it with a gas water heater unless you are willing and able to install new electrical.

Use the comment column in the following table to list when each step was completed, who complete the step, and/or any other special circumstances that might have arisen during the completion of that step.

Step	Comment
1. Turn off the gas or electricity to the heater.	
2. Drain the heater. Opening the T&P value will let air into the system.	
3. Disconnect the water lines.	
4. Disconnect the fuel source to heater.	
5. Put the new heater in its location by "walking" it or using an appliance cart, dolly, or hand truck. Position the new heater so the piping—particularly gas vent piping—will reach most easily.	
6. If you removed the shutoff valve, replace it.	
7. Install the water lines and pressure relief line.	
8. Reconnect the fuel source line.	

Note: Follow the manufacturer's instructions for specific step-by-step lists on how to install specific water heaters (i.e., gas or electric).

Instructor's Response: Turn page

Instructor's Response

JOB SHEET 3

Gas Water Heater

Name: _____   **Date:** _____

After completing this job sheet, the student should be able to identify the gas pipe connection to a water heater.

Write in labels/notes to correspond with arrows in the following illustration.

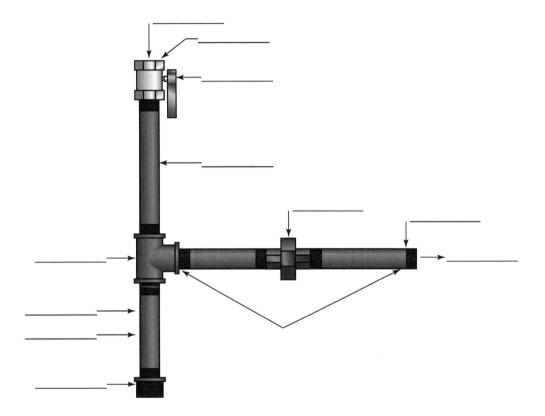

Instructor's Response

OBJECTIVES

After completing this chapter, the student should be able to:

- Understand hot water heater controls.
- Understand why controls and devices are important.
- Respect safety devices.

KEY TERMS

Burner assembly

High limit

Leg of power

Non-simultaneous

Pilot

T&P

Terminal

Thermocouple

Thermostat

INTRODUCTION

Hot water is used extensively in residential, commercial, and industrial plumbing applications. In residential applications, it is used for cooking and cleaning, washing clothes, washing dishes, and bathing. In commercial applications it is used for cooking, cleaning, and sanitizing. In industrial applications it is used for numerous manufacturing processes. However, regardless of the use and the application the means by which hot water is created is still basically the same: A piece of equipment is used to heat the water and that equipment must be self-regulating and the heating cycles must be automatic.

REVIEW QUESTIONS

Completion

1. _____ is a single energized wire from an electrical source to a heating device.

2. _____ is the abbreviation for temperature and pressure. It describes a dual-purpose relief valve that activates (opens) when it senses excessive temperature and pressure in a water heater.

3. _____ is the location where electrical wires connect to a device.

4. _____ is a heat-sensing device.

5. _____ is the flame that ignites incoming gas to a burner assembly.

6. _____ describes a heating cycle of an electric water heater that only has one heating element energized at a time.

7. _____ is the area of a gas water heater where a flame is ignited to heat water.

8. _____ is a safety device that senses excessive water temperature and disconnects the electrical circuit to stop a heating cycle.

9. _____ is a temperature-regulating device that senses high and low temperatures.

JOB SHEET 1

Identify the Parts of a Gas Water Heater Regulator

Name: _____   Date: _____

After completing this job sheet, the student should be able to identify the parts of a gas water heater regulator.

Write in labels to correspond with arrows in the following illustrations.

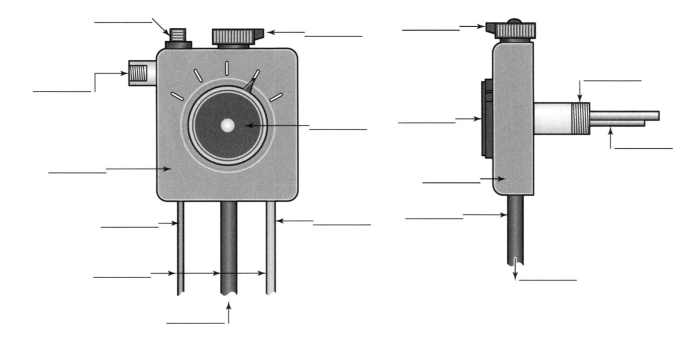

Instructor's Response

JOB SHEET 2

Identify the Parts of an Electric Residential Water Heater

Name: _____   Date: _____

After completing this job sheet, the student should be able to identify the parts of an electric residential water heater.

Write in labels to correspond with arrows in the following illustrations.

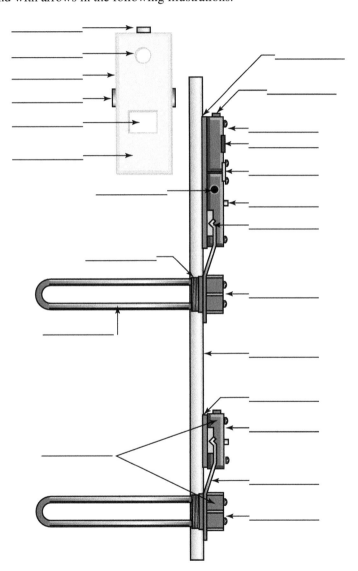

Instructor's Response

SECTION 5

Drainage, Waste, and Vent Systems

14 Drainage, Waste, and Vent System Identification

OBJECTIVES

After completing this chapter, the student should be able to:

- Identify sewage disposal locations.
- Understand and identify the different segments of a drainage, waste, and vent system.

KEY TERMS

Air admittance valve

Branch interval

Cleanout

Invert

Manhole

Ordinances

INTRODUCTION

Supplying water to a facility is only a portion of the plumbing system. Proper waste removal is just as important. However, currently there are more plumbing codes regulating drain, waste, and vent systems than any other plumbing system. This is partially due to the fact that proper waste handling and proper waste disposal are major health concerns.

REVIEW QUESTIONS

Completion

1. _____ is an access point into a drain to clean an obstruction in the drain.

2. _____ is a structure installed within a sewer system that provides a cleanout location.

3. _____ is a venting device installed in a drainage system. It only allows air to enter the system and does not allow sewer gas to escape.

4. _____ is the location where a drainage branch connects to a waste stack.

5. _____ is the bottom of a drain pipe.

6. _____ are local codes enacted by local governments and are only enforceable in that municipal jurisdiction.

JOB SHEET **1**

Major Segments of a DWV

Name: _____   Date: _____

After completing this job sheet, the student should be able to identify the major segments of a DWV system.

Insert missing abbreviations and descriptions in the following table.

Segment	Abbreviation	Description
Building sewer		Exterior piping conveying sewage and wastewater from the BD to the point of disposal
Waste stack		
	BD	
Vent stack		

Instructor's Response

JOB SHEET 2

Minor Segments of a DWV

Name: _____ Date: _____

After completing this job sheet, the student should be able to identify the minor segments of a DWV system.

Insert missing abbreviations and descriptions in the following table.

Segment	Abbreviation	Description
Branch vent		
Circuit vent		
Fixture branch		
Fixture drain		
Horizontal branch		
	IV	
	LV	
	RV	
		Piping arrangement that over-sizes the pipe to serve as a drain and a vent and is the only vent that conveys wastewater

Instructor's Response

OBJECTIVES

After completing this chapter, the student should be able to:

- Identify traps, floor drains, interceptors, and carriers.

- Understand the importance of the drainage, waste, and vent layout process.

- Understand the basics of drainage, waste, and vent installations.

KEY TERMS

Area drain

Batter board

Chase

Floor drain

Floor joist

Floor sink

Grade

Interceptor

Siphon

Trap

Trap primer

INTRODUCTION

Before a drain, waste, and vent system can be installed, there are several design consideration that must be addressed. These considerations include placement of appliances and fixtures, plumbing codes, and other utilities and systems.

REVIEW QUESTIONS

Completion

1. _____ is a floor drain serving locations such as a driveway, garage, or a patio.

2. _____ is a space provided behind a wall to install pipe and carriers.

3. _____ are horizontal structural support for a floor.

4. _____ is used to describe a pipe installed at a slight angle.

5. _____ is a device installed in a drainage system to intercept undesirable items or fluids from entering the sewer system, such as grease, oil, and sand.

6. _____ is a type of floor drain that has capacity to receive discharge from equipment such as a commercial dishwasher.

7. _____ is a drain that serves as a drainage inlet for a floor, and can be used to receive discharge from small pieces of equipment.

8. _____ are two vertical wood boards and one horizontal wood board assembled on a job site to anchor strings for layout purposes.

9. _____ is an action of water being pulled from a trap into a drain.

10. _____ is a device installed to provide water to a trap to replenish a trap seal.

11. _____ is a protective device (fitting) installed at each fixture to minimize sewer gas entering an occupied space.

Drain, Waste, and Vent Considerations

Name: _____   Date: _____

After completing this job sheet, the student should be able to identify some of the considerations for laying out a drain, waste, and vent system.

Explain why each consideration listed in the following table is important.

Consideration	Comment
Fixture location	
Type of specific fixture	
Codes	
Locations of studs, joist, or other structural supports	
Physical size of holes and floor openings	
Coordination with other building trades	

Instructor's Response

JOB SHEET 2

Calculating Fall per Foot Formula

Name: _____   Date: _____

After completing this job sheet, the student should be able to calculate the amount of fall necessary of a length of pipe.

Find the fall per foot for each of the pipes listed in the following table.

Pipe length	Slope per foot	Answer
255 feet	1/4"	
35 feet	1/8"	
137.5 feet	1/4"	

Instructor's Response

16 Drainage, Waste, and Vent Installations

OBJECTIVES

After completing this chapter, the student should be able to:

- Identify the important steps in installing a drainage, waste, and vent system.
- Recognize that many installation procedures are based on a specific project.
- Test a drainage, waste, and vent system.

KEY TERMS

Band iron

Fire-rated wall

Flood-level rim

Floor joists

Footing

Load bearing

Stud-guard

VTR

Wall studs

INTRODUCTION

The installation of a drainage, waste, and vent (DWV) system is often completed in several phases in residential and commercial construction. When the drainage is to be installed below a slab, the piping is typically done during the first phase of the construction, called the underground.

REVIEW QUESTIONS

Completion

1. _____ describes a wall or other structural component that is bearing the weight of a portion of a structure.

2. _____ is installed where a pipe passes close to the edge of a stud or joist to protect the pipe from nails and screws.

3. _____ is used on a blueprint to illustrate that a vent pipe is terminated above a roof.

4. _____ is the horizontal structural board supporting a floor.

5. _____ is a concrete structural component of a building to support walls and columns.

6. _____ is the height at which a fixture would overflow onto the floor, such as a countertop.

7. _____ is a wall constructed to slow the spread of fire.

8. _____ is sold in rolls and cut to length to support piping.

9. _____ is a vertical wall support member that is either load-bearing or non-load-bearing.

JOB SHEET 1

Working with PVC Pipe

Name: _____   Date: _____

After completing this job sheet, the student should be able to cut and glue PVC pipe.

Use the comment column in the following table to list when each step was completed, who complete the step, and/or any other special circumstances that might have arisen during the completion of that step.

Step	Comment
1. Measure and cut the pipe to the desired length using a saw or tubing cutter.	
2. Remove burrs with a file, sand paper, or deburring tool.	
3. Apply PVC cleaner (purple primer) to pipe and fitting socket as recommended by manufacturer.	
4. Apply PVC cleaner to inside of fitting socket.	
5. Apply glue to pipe end that is to be glued.	
6. Apply glue to fitting socket.	
7. Insert pipe end into fitting socket and twist at least one-quarter turn while holding the joint together for about one minute.	

Instructor's Response

Roof Drain Systems and Sewer and Drain Cleaning

OBJECTIVES

After completing this chapter, the student should be able to:

- Identify the segments of a roof drain system.
- Understand the scope of a storm drain system.
- Identify the basics of clearing sewer and drain stoppages.

KEY TERMS

Conductor

Downspout

Grout

Leader

Parapet

Peak rainfall periods

Sweating

INTRODUCTION

One of the most frequent types of calls that a plumber will receive is either a clogged lavatory or toilet. These situations usually can be resolved without the assistance of a professional by following a few basic steps.

REVIEW QUESTIONS

Completion

1. _____ is the placement of a grout mixture under and on top of a catch basin grate to seal its connection with the catch basin.

2. _____ is located on the inside of a building. It conveys storm water vertically from upper horizontal storm drain piping to a building storm drain.

3. _____ describes the process of condensation forming on the exterior of a pipe.

4. _____ is a vertical pipe or square tube conveying water from a gutter to the ground.

5. _____ is located on the outside of a building. It conveys storm water vertically from upper horizontal storm drain piping to a building storm sewer.

6. _____ is a vertical wall constructed on a roof edge.

7. _____ is the amount of rainfall projected in one hour based on certain historical data over 10, 25, or 100 years.

JOB SHEET 1

Unclogging a Kitchen Sink or Lavatory Using a Plunger

Name: _____   Date: _____

After completing this job sheet, the student should be able to unclog a kitchen sink using a plunger.

Use the comment column in the following table to list when each step was completed, who completed the step, and/or any other special circumstances that might have arisen during the completion of that step.

Step	Comment
1. Remove the sink's strainer or plug.	
2. If the sink does not already have water in it, fill the sink with water halfway.	
3. Carefully place the plunger's tuber globe over the drain ensuring that the entire drain opening is covered by the plunger.	
4. Using forceful strokes, plunge the sink drain at least 15 times before removing the plunger to see if the sink will drain.	
5. If the sink is still clogged, repeat Steps 3 and 4.	
6. If the sink cannot be unclogged with a plunger: Remove the trap, clean it out, and reassemble it.	
7. Once the sink has been unclogged, run hot water down the drain for several minutes. This will help cleanout anything remaining in the system.	

Instructor's Response

JOB SHEET 2

Unclogging a Toilet Using a Plunger

Name: _____   Date: _____

After completing this job sheet, the student should be able to unclog a toilet using a plunger.

Use the comment column in the following table to list when each step was completed, who completed the step, and/or any other special circumstances that might have arisen during the completion of that step.

Step	Comment
1. Insert the plunger into the toilet bowl fully covering the drain opening.	
2. Pushing down and pulling up on the handle of the plunger vigorously, plunge the toilet 15 to 20 times.	
3. Lift the plunger from the drain opening to see if the toilet drains.	
4. If toilet is still clogged, repeat Steps 1 through 3.	
5. If the toilet cannot be unclogged with a plunger: Remove the toilet, turn it upside down, remove the obstruction, and reassemble.	

Instructor's Response

JOB SHEET 3

Unclogging a Toilet Using an Auger

Name: _____   Date: _____

After completing this job sheet, the student should be able to unclog a toilet using an auger.

Use the comment column in the following table to list when each step was completed, who completed the step, and/or any other special circumstances that might have arisen during the completion of that step.

Step	Comment
1. Loosen the setscrews on the auger and push the cable into the drain moving it back and forth until the clog is reached.	
2. Tighten the setscrews on the toilet auger.	
3. While pushing on the toilet auger, crank the auger clockwise until the obstruction is cleared.	
4. Remove the toilet auger from toilet.	
5. Test the flow of the toilet by flushing it several times.	

Instructor's Response

SECTION 6

System Sizing

Sizing Drainage, Waste, and Vent Systems

OBJECTIVES

After completing this chapter, the student should be able to:

- Understand sizing methods of various segments of a drainage, waste, and vent system.

- Interpret codebook sizing tables.

- Understand sizing approaches on a job site.

- Recognize that some minimum pipe-size codes override sizing tables.

KEY TERMS

Arbitrarily chosen

Continuous flow

Drainage fixture unit

Indirect waste

Semi-continuous flow

INTRODUCTION

One of the most difficult aspects of sizing a drainage, waste, and vent (DWV) system is to know various over-riding codes based on the location of an installation. Segment identification is vital in locating the correct sizing information, but many codes relate to where the segments are installed as opposed to its definition. Some charts in a codebook provide numerous options based on a location of an installation and a plumber must correctly use the information provided in a codebook.

DWV SIZING

Layout for a DWV system is typically the first process completed on a project and other piping systems are installed around the DWV system. Pipe and fitting sizes are larger than other piping systems in a residential building and code regulations, specific fixture locations, and available space to install piping necessitate its installation or layout to be an initial activity. Layout is based on a particular objective, various codes, and sizing requirements. Sizing of piping is initially based on theory, but health issues combined with practical considerations determine how codes regulate sizing of a DWV system. Sizing is one key area a plumber must comprehend to become licensed.

Theoretical methods that determine sizing are based on the rate of flow that a fixture discharges wastewater into a drainage system. The measuring factor is known as a drainage fixture unit (dfu) and the average plumber does not perform the calculations on a job site. A plumber must use the known dfu values of fixtures to locate correct sizing information in a codebook. Most dfu values in a codebook are based on flow rates of certain fixtures, pipe sizes, and the dfu load effects on a DWV system.

REVIEW QUESTIONS

Completion

1. _____ is a pipe discharging liquid from equipment into a drainage system and terminating with an air gap above a floor drain or fixture.

2. Fixture units indicating the theoretic values are adjusted for practical use and are _____.

3. _____ water flow into a drainage system that is introduced is the method used to assign a value for a drainage automatically, or that produces more than intermittent and less than continuous flow, such as from a piece of equipment.

4. _____ is a factor used to size pipe. Dfu equals the flow measured in gallons per minute or liters per second into a drainage system from a plumbing fixture.

5. _____ is water flow into a drainage system that is introduced either automatically or constantly, such as from a piece of equipment.

JOB SHEET 1

Typical Vent Sizing Table

Name: _____   Date: _____

After completing this job sheet, the student should be able to read a typical vent sizing table.

		Maximum Developed Length of Vent (in Feet)				
Stack Size	DFU	1 1/4" Pipe	1 1/2" Pipe	2" Pipe	3" Pipe	4" Pipe
1 1/4	2	30	0	0	0	0
1 1/2	8	50	150	0	0	0
1 1/2	10	30	100	0	0	0
2	12	30	75	200	0	0
2	20	26	50	150	0	0
1 1/2	42	0	30	100	0	0
3	10	0	42	150	1040	0
3	21	0	32	110	810	0
3	53	0	27	94	680	0
3	102	0	25	86	620	0
4	43	0	0	35	250	980
4	140	0	0	27	200	750
4	320	0	0	23	170	640
4	540	0	0	21	150	580

Refer to the preceding table to determine the following:

   a. The maximum length in feet of a vent that is to service a 1 1/2" stack with 8 dfu if the vent size is to be 1 1/2".

   b. The maximum length in feet of a vent that is to service a 3" stack with 10 dfu if the vent size is to be 3".

c.   The maximum length in feet of a vent that is to service a 1 1/4" stack with 2 dfu if the vent size is to be 1 1/4".

d.   The maximum length in feet of a vent that is to service a 1 1/2" stack with 10 dfu if the vent size is to be 1 1/2".

Instructor's Response

OBJECTIVES

After completing this chapter, the student should be able to:

- Understand sizing methods of natural gas and water supply systems.

- Size a water heater.

- Understand different sizing approaches.

KEY TERMS

Friction loss

Hardness

Pressure drop

Pressure-loss methods

Velocity methods

INTRODUCTION

The gas supply pipe arrangement is fairly typical with most residential water heaters and the gas regulator (gas valve) has a 1/2" female threaded connection. Most codes dictate that a piping configuration utilizing what is known as a *drip leg* be installed to collect any small particles or moisture present in the gas supply system. This configuration is assembled using Schedule 40 black steel pipe nipples and black malleable pipe fitting. Flexible metallic connectors designed for this purpose are often installed and are allowable by local codes. A gas cock or other approved isolation valves are installed near the water heater. The venting requirements of a gas water heater is dictated by code, type of heater, and job site conditions.

REVIEW QUESTIONS

Completion

1. _____ is the total mineral content of water.

2. _____ is when pressure decreases due to friction as gas or water flows through piping.

3. _____ is a system for sizing water pipe based on velocity of the flow of water through a system.

4. _____ is the loss of pressure from friction or obstructions from pipe, valve fittings, and devices.

5. _____ is a system for sizing water pipe based on pressure loss through the system.

JOB SHEET 1

Gas Water Heater Selection

Name: _____   Date: _____

After completing this job sheet, the student should be able to follow the correct sequence for selecting a gas water heater.*

		No flow	Temperature rise		Gallons per hour
Model Number	BTU	First Hour	100°	90°	60°
1234	35,000	60	40	45	55
2345	40,000	75	55	60	70
3456	45,000	80	50	65	75
4567	50,000	85	60	70	80
5678	55,000	95	70	80	90
6789	60,000	100	80	85	95

* Values are for lesson purpose and not real values.

Note: If the temperature rise is not known when sizing a water heater, sizing calculations typically utilize 100-degree temperature rise.

Complete the following table to calculate total GPH, and then refer to the preceding chart to determine the correct water heater.

	Number	GPH/Unit	Subtotal GPH
Bath	1	20	
Kitchen	1	20	
Laundry Room	1	30	
		Total GPH	

The correct water heater is model number _____.

Instructor's Response